我的人生解答书

The

ANSWERS

of

MY LIFE

命运之门 · 为你洞开

The

ANSWERS

of

MY LIFE

使用手册

1. 双手捧起书，封面贴胸而放。
2. 默想你心中的问题，尽量是简单的问答句式。
   例如“这条路如何走下去？”“我该放弃她吗？”
3. 闭上眼睛，默想问题三遍。
4. 深呼吸，然后翻开书，睁眼看这一页的答案。
5. 希望对你的人生有帮助。

——你所有的问题，这本书都能替你解答——

The
ANSWERS
of
MY LIFE

有风险，但也有机会

The

ANSWERS

of

MY LIFE

找个人给你点意见

The

ANSWERS

of

MY LIFE

算了吧

The ANSWERS of MY LIFE

请教你妈妈

The ANSWERS *of* MY LIFE

当然咯

The ANSWERS *of* MY LIFE

谁说得准呢，先观望着

The
ANSWERS
of
MY LIFE

千万别傻

The
ANSWERS
of
MY LIFE

保持你的好奇心，去挖掘真相

The
ANSWERS
of
MY LIFE

把心揣怀里

The

ANSWERS

of

MY LIFE

答案在镜子里

The
ANSWERS
of
MY LIFE

NO

The ANSWERS *of* MY LIFE

这事儿不靠谱

The

ANSWERS

of

MY LIFE

天上要掉馅饼了

The

ANSWERS

of

MY LIFE

有好运

The

ANSWERS

of

MY LIFE

要有耐心

The ANSWERS *of* MY LIFE

你需要知道真相

The

ANSWERS

of

MY LIFE

还有另一种情况

The
ANSWERS
of
MY LIFE

观望

The

ANSWERS

of

MY LIFE

别让它影响到你

The

ANSWERS

of

MY LIFE

YES

The

ANSWERS

of

MY LIFE

信任

The ANSWERS *of* MY LIFE

列个清单

The ANSWERS of MY LIFE

时机不对

The

ANSWERS

of

MY LIFE

照你想做的那样去做

The

ANSWERS

of

MY LIFE

量力而行

The

ANSWERS

of

MY LIFE

但行好事，莫问前程

The
ANSWERS
of
MY LIFE

抛弃首选方案

The

ANSWERS

of

MY LIFE

走容易走的路

The

ANSWERS

of

MY LIFE

最佳方案不一定可行

The

ANSWERS

of

MY LIFE

NO ZUO NO DIE

The

ANSWERS

of

MY LIFE

试试卖萌

The

ANSWERS

of

MY LIFE

借助他人的经验

The

ANSWERS

of

MY LIFE

再多考虑

The

ANSWERS

of

MY LIFE

注意细节

The

ANSWERS

of

MY LIFE

说出来吧

The

ANSWERS

of

MY LIFE

不要犹豫

The

ANSWERS

of

MY LIFE

机会稍纵即逝

The

ANSWERS

of

MY LIFE

制订一个新计划

The ANSWERS *of* MY LIFE

GO

The

ANSWERS

of

MY LIFE

谁都不能保证

The ANSWERS *of* MY LIFE

情况很快会发生变化

The

ANSWERS

of

MY LIFE

不要陷得太深

The

ANSWERS

of

MY LIFE

转移你的注意力

The
ANSWERS
of
MY LIFE

至关重要

The
ANSWERS
of
MY LIFE

告诉自己什么是最重要的

The
ANSWERS
of
MY LIFE

为什么不

The

ANSWERS

of

MY LIFE

别傻等了

The

ANSWERS

of

MY LIFE

不要忘记

The

ANSWERS

of

MY LIFE

WHY NOT

The

ANSWERS

of

MY LIFE

去解决

The

ANSWERS

of

MY LIFE

寻找更多的选择

The

ANSWERS

of

MY LIFE

上帝为你关一扇门，
必定会再为你开一扇窗

The

ANSWERS

of

MY LIFE

随波逐流未必是好事

The

ANSWERS

of

MY LIFE

问天问大地，不如问问自己

The

ANSWERS

of

MY LIFE

你就是答案

The

ANSWERS

of

MY LIFE

去争取机会

The

ANSWERS

of

MY LIFE

改变不了世界，改变自己

The

ANSWERS

of

MY LIFE

主动一点，人生会大不相同

The
ANSWERS
of
MY LIFE

学会妥协

The

ANSWERS

of

MY LIFE

掌握更多信息

The

ANSWERS

of

MY LIFE

相信你最初的想法

The

ANSWERS

of

MY LIFE

勿忘初心，方得始终

The ANSWERS *of* MY LIFE

扫清障碍

The

ANSWERS

of

MY LIFE

把重心放在工作 / 学习上

The

ANSWERS

of

MY LIFE

培养一项新的爱好

The

ANSWERS

of

MY LIFE

对他人慷慨

The
ANSWERS
of
MY LIFE

不妨赌一把

The
ANSWERS
of
MY LIFE

去做其他的事情

The
ANSWERS
of
MY LIFE

观察形势

The
ANSWERS
of
MY LIFE

休息，休息一会儿

The ANSWERS *of* MY LIFE

这是你最后的机会

The
ANSWERS
of
MY LIFE

再考虑一下

The

ANSWERS

of

MY LIFE

并不明智

The
ANSWERS
of
MY LIFE

等待更好的

The

ANSWERS

of

MY LIFE

很快能解决

The
ANSWERS
of
MY LIFE

重要

The ANSWERS *of* MY LIFE

是的

The

ANSWERS

of

MY LIFE

采取行动

The

ANSWERS

of

MY LIFE

去做

The ANSWERS *of* MY LIFE

不要过火

The
ANSWERS
of
MY LIFE

事情开始变得有趣了

The

ANSWERS

of

MY LIFE

保存你的实力

The ANSWERS *of* MY LIFE

这是一定的

The

ANSWERS

of

MY LIFE

不确定因素有点多

The ANSWERS *of* MY LIFE

结果不错

The

ANSWERS

of

MY LIFE

你可能不得不放弃其他东西

The

ANSWERS

of

MY LIFE

不要忧虑

The
ANSWERS
of
MY LIFE

不需要

The

ANSWERS

of

MY LIFE

去倾诉

The

ANSWERS

of

MY LIFE

告诉别人这对你意味着什么

The
ANSWERS
of
MY LIFE

无论你做何种选择，结果都是对的

The

ANSWERS

of

MY LIFE

保持头脑清醒

The

ANSWERS

of

MY LIFE

制订计划

The

ANSWERS

of

MY LIFE

很麻烦

The ANSWERS *of* MY LIFE

克服困难

The

ANSWERS

of

MY LIFE

实际一点

The ANSWERS *of* MY LIFE

你需要一点帮助

The

ANSWERS

of

MY LIFE

协作

The

ANSWERS

of

MY LIFE

寻找更多的选择

The

ANSWERS

of

MY LIFE

负责

The

ANSWERS

of

MY LIFE

阻止

The
ANSWERS
of
MY LIFE

你必须现在就行动

The

ANSWERS

of

MY LIFE

遵守规则

The

ANSWERS

of

MY LIFE

坚持

The

ANSWERS

of

MY LIFE

你不会失望的

The
ANSWERS
of
MY LIFE

需要花费点时间

The

ANSWERS

of

MY LIFE

不要迫于压力而改变初衷

The
ANSWERS
of
MY LIFE

显而易见

The

ANSWERS

of

MY LIFE

不要忽略身边的人

The
ANSWERS
of
MY LIFE

抗拒

The

ANSWERS

of

MY LIFE

不值得斗争

The

ANSWERS

of

MY LIFE

玩得开心就好

The ANSWERS *of* MY LIFE

毋庸置疑

The
ANSWERS
of
MY LIFE

你也许会失望

The

ANSWERS

of

MY LIFE

去改变

The
ANSWERS
of
MY LIFE

一个强有力的承诺将会换回更好的结果

The

ANSWERS

of

MY LIFE

也许有更好的解决方案

The

ANSWERS

of

MY LIFE

不要害怕

The

ANSWERS

of

MY LIFE

想法太多，选择太少

The

ANSWERS

of

MY LIFE

是的

The ANSWERS *of* MY LIFE

一笑而过

The

ANSWERS

of

MY LIFE

取决于你的选择

The

ANSWERS

of

MY LIFE

随他去

The
ANSWERS
of
MY LIFE

你需要考虑其他方面

The

ANSWERS

of

MY LIFE

一年后就不那么重要了

The
ANSWERS
of
MY LIFE

醒醒吧，别做梦了

The
ANSWERS
of
MY LIFE

意义非凡

The

ANSWERS

of

MY LIFE

默数十秒再问我

The
ANSWERS
of
MY LIFE

去行动

The
ANSWERS
of
MY LIFE

发挥你的想象力

The ANSWERS *of* MY LIFE

对的

The

ANSWERS

of

MY LIFE

为了确保最好的决定，保持冷静

The

ANSWERS

of

MY LIFE

等待

The

ANSWERS

of

MY LIFE

你必须弥补这个缺点

The

ANSWERS

of

MY LIFE

你会后悔的

The

ANSWERS

of

MY LIFE

毫无疑问

The
ANSWERS
of
MY LIFE

当然

The ANSWERS *of* MY LIFE

现在比以往任何时候的情况都要好

The
ANSWERS
of
MY LIFE

相信你的直觉

The
ANSWERS
of
MY LIFE

这是一个机会

The

ANSWERS

of

MY LIFE

去问你爸爸

The

ANSWERS

of

MY LIFE

从来没有

The
ANSWERS
of
MY LIFE

寻找一个指路人

The

ANSWERS

of

MY LIFE

去尝试

The

ANSWERS

of

MY LIFE

没有

The

ANSWERS

of

MY LIFE

错的

The
ANSWERS
of
MY LIFE

别不自量力

The

ANSWERS

of

MY LIFE

荒谬

The

ANSWERS

of

MY LIFE

不赌

The

ANSWERS

of

MY LIFE

不值得冒险

The

ANSWERS

of

MY LIFE

不妥协

The
ANSWERS
of
MY LIFE

关注你的家庭生活

The
ANSWERS
of
MY LIFE

肯定

The

ANSWERS

of

MY LIFE

不可预测

The

ANSWERS

of

MY LIFE

绝对不

The ANSWERS *of* MY LIFE

我确定

The

ANSWERS

of

MY LIFE

尽早完成

The ANSWERS *of* MY LIFE

令人期待的事情马上要发生

The

ANSWERS

of

MY LIFE

你需要适应

The

ANSWERS

of

MY LIFE

表示怀疑

The
ANSWERS
of
MY LIFE

它会带来好运

The ANSWERS *of* MY LIFE

要有耐心

The

ANSWERS

of

MY LIFE

看看会发生什么

The

ANSWERS

of

MY LIFE

记录下来

The

ANSWERS

of

MY LIFE

不宜在这个时候

The

ANSWERS

of

MY LIFE

决定了就去做

The

ANSWERS

of

MY LIFE

别要求太多

The
ANSWERS
of
MY LIFE

放弃第一个方案

The
ANSWERS
of
MY LIFE

HOLD 不住

The ANSWERS *of* MY LIFE

谨慎小心

The
ANSWERS
of
MY LIFE

注意细节

The

ANSWERS

of

MY LIFE

注意身后

The ANSWERS *of* MY LIFE

不要犹豫

The

ANSWERS

of

MY LIFE

继续前进

The
ANSWERS
of
MY LIFE

情况很快会发生改变

The

ANSWERS

of

MY LIFE

不要被情绪左右

The
ANSWERS
of
MY LIFE

转移注意力

The

ANSWERS

of

MY LIFE

着眼未来

The
ANSWERS
of
MY LIFE

问自己什么是最重要的

The

ANSWERS

of

MY LIFE

不要等了

The ANSWERS *of* MY LIFE

保持乐观

The

ANSWERS

of

MY LIFE

没有更好的选择

The ANSWERS *of* MY LIFE

列出原因

The

ANSWERS

of

MY LIFE

抓住机会

The

ANSWERS

of

MY LIFE

改变自己

The
ANSWERS
of
MY LIFE

你需要主动

The ANSWERS *of* MY LIFE

妥协

The
ANSWERS
of
MY LIFE

有比这更重要的东西

The

ANSWERS

of

MY LIFE

你需要掌握更多的信息

The

ANSWERS

of

MY LIFE

删除记忆

The

ANSWERS

of

MY LIFE

专注于你的工作

The
ANSWERS
of
MY LIFE

做能让你快乐的那个决定

The
ANSWERS
of
MY LIFE

你需要考虑其他方面

The

ANSWERS

of

MY LIFE

你会后悔的

The

ANSWERS

of

MY LIFE

毫无疑问

The

ANSWERS

of

MY LIFE

当然

The ANSWERS *of* MY LIFE

相信自己的直觉

The

ANSWERS

of

MY LIFE

这是一个机会

The
ANSWERS
of
MY LIFE

形勢不明

The

ANSWERS

of

MY LIFE

先让自己休息

The

ANSWERS

of

MY LIFE

重新考虑

The

ANSWERS

of

MY LIFE

不明智

The

ANSWERS

of

MY LIFE

抓住机会

The
ANSWERS
of
MY LIFE

等待机会

The
ANSWERS
of
MY LIFE

不要做得太过分

The

ANSWERS

of

MY LIFE

保持现状

The

ANSWERS

of

MY LIFE

不要忧虑

The
ANSWERS
of
MY LIFE

有意料之外的事会发生，不妨等待

The

ANSWERS

of

MY LIFE

你会失望的

The
ANSWERS
of
MY LIFE

花更多的时间来决定

The ANSWERS *of* MY LIFE

你开心就好

The ANSWERS *of* MY LIFE

有风险，但也有机会

The

ANSWERS

of

MY LIFE

找个人给你点意见

The

ANSWERS

of

MY LIFE

算了吧

The ANSWERS *of* MY LIFE

请教你妈妈

The ANSWERS *of* MY LIFE

当然咯

The

ANSWERS

of

MY LIFE

谁说得准呢，先观望着

The

ANSWERS

of

MY LIFE

千万别傻

The ANSWERS *of* MY LIFE

保持你的好奇心，去挖掘真相

The
ANSWERS
of
MY LIFE

把心揣怀里

The ANSWERS *of* MY LIFE

答案在镜子里

The

ANSWERS

of

MY LIFE

NO

The ANSWERS *of* MY LIFE

这事儿不靠谱

The

ANSWERS

of

MY LIFE

天上要掉馅饼了

The

ANSWERS

of

MY LIFE

有好运

The
ANSWERS
of
MY LIFE

要有耐心

The ANSWERS *of* MY LIFE

你需要知道真相

The

ANSWERS

of

MY LIFE

还有另一种情况

The ANSWERS *of* MY LIFE

观望

The

ANSWERS

of

MY LIFE

别让它影响到你

The ANSWERS *of* MY LIFE

YES

The
ANSWERS
of
MY LIFE

信任

The

ANSWERS

of

MY LIFE

列个清单

The

ANSWERS

of

MY LIFE

时机不对

The ANSWERS *of* MY LIFE

照你想做的那样去做

The

ANSWERS

of

MY LIFE

量力而行

The

ANSWERS

of

MY LIFE

但行好事，莫问前程

The

ANSWERS

of

MY LIFE

抛弃首选方案

The
ANSWERS
of
MY LIFE

走容易走的路

The

ANSWERS

of

MY LIFE

最佳方案不一定可行

The

ANSWERS

of

MY LIFE

NO ZUO NO DIE

The ANSWERS *of* MY LIFE

试试卖萌

The ANSWERS *of* MY LIFE

借助他人的经验

The

ANSWERS

of

MY LIFE

再多考虑

The

ANSWERS

of

MY LIFE

注意细节

The

ANSWERS

of

MY LIFE

说出来吧

The ANSWERS *of* MY LIFE

不要犹豫

The

ANSWERS

of

MY LIFE

机会稍纵即逝

The
ANSWERS
of
MY LIFE

制订一个新计划

The

ANSWERS

of

MY LIFE

GO

The ANSWERS *of* MY LIFE

谁都不能保证

The

ANSWERS

of

MY LIFE

情况很快会发生变化

The

ANSWERS

of

MY LIFE

不要陷得太深

The

ANSWERS

of

MY LIFE

转移你的注意力

The

ANSWERS

of

MY LIFE

至关重要

The ANSWERS *of* MY LIFE

告诉自己什么是最重要的

The

ANSWERS

of

MY LIFE

为什么不

The

ANSWERS

of

MY LIFE

别傻等了

The

ANSWERS

of

MY LIFE

不要忘记

The
ANSWERS
of
MY LIFE

WHY NOT

The ANSWERS *of* MY LIFE

去解决

The

ANSWERS

of

MY LIFE

寻找更多的选择

The
ANSWERS
of
MY LIFE

上帝为你关一扇门，
必定会再为你开一扇窗

The

ANSWERS

of

MY LIFE

随波逐流未必是好事

The

ANSWERS

of

MY LIFE

问天问大地，不如问问自己

The
ANSWERS
of
MY LIFE

你就是答案

The

ANSWERS

of

MY LIFE

去争取机会

The ANSWERS *of* MY LIFE

改变不了世界，改变自己

The
ANSWERS
of
MY LIFE

主动一点，人生会大不相同

The

ANSWERS

of

MY LIFE

学会妥协

The ANSWERS *of* MY LIFE

掌握更多信息

The ANSWERS *of* MY LIFE

相信你最初的想法

The

ANSWERS

of

MY LIFE

勿忘初心，方得始终

The

ANSWERS

of

MY LIFE

扫清障碍

The ANSWERS *of* MY LIFE

把重心放在工作/学习上

The

ANSWERS

of

MY LIFE

培养一项新的爱好

The

ANSWERS

of

MY LIFE

对他人慷慨

The ANSWERS *of* MY LIFE

不妨赌一把

The ANSWERS *of* MY LIFE

去做其他的事情

The

ANSWERS

of

MY LIFE

观察形势

The

ANSWERS

of

MY LIFE

休息，休息一会儿

The
ANSWERS
of
MY LIFE

这是你最后的机会

The

ANSWERS

of

MY LIFE

再考虑一下

The

ANSWERS

of

MY LIFE

并不明智

The

ANSWERS

of

MY LIFE

等待更好的

The

ANSWERS

of

MY LIFE

很快能解决

The

ANSWERS

of

MY LIFE

重要

The ANSWERS *of* MY LIFE

是的

The

ANSWERS

of

MY LIFE

采取行动

The

ANSWERS

of

MY LIFE

去做

The ANSWERS *of* MY LIFE

不要过火

The

ANSWERS

of

MY LIFE

事情开始变得有趣了

The

ANSWERS

of

MY LIFE

保存你的实力

The

ANSWERS

of

MY LIFE

这是一定的

The ANSWERS *of* MY LIFE

不确定因素有点多

The
ANSWERS
of
MY LIFE

结果不错

The ANSWERS *of* MY LIFE

你可能不得不放弃其他东西

The

ANSWERS

of

MY LIFE

不要忧虑

The

ANSWERS

of

MY LIFE

不需要

The

ANSWERS

of

MY LIFE

去倾诉

The

ANSWERS

of

MY LIFE

告诉别人这对你意味着什么

The

ANSWERS

of

MY LIFE

无论你做何种选择，结果都是对的

The
ANSWERS
of
MY LIFE

保持头脑清醒

The ANSWERS *of* MY LIFE

制订计划

The
ANSWERS
of
MY LIFE

很麻烦

The

ANSWERS

of

MY LIFE

克服困难

The

ANSWERS

of

MY LIFE

实际一点

The

ANSWERS

of

MY LIFE

你需要一点帮助

The

ANSWERS

of

MY LIFE

协作

The
ANSWERS
of
MY LIFE

寻找更多的选择

The ANSWERS *of* MY LIFE

负责

The ANSWERS *of* MY LIFE

阻止

The

ANSWERS

of

MY LIFE

你必须现在就行动

The

ANSWERS

of

MY LIFE

遵守规则

The ANSWERS *of* MY LIFE

坚持

The ANSWERS *of* MY LIFE

你不会失望的

The

ANSWERS

of

MY LIFE

需要花费点时间

The

ANSWERS

of

MY LIFE

不要迫于压力而改变初衷

The

ANSWERS

of

MY LIFE

显而易见

The

ANSWERS

of

MY LIFE

不要忽略身边的人

The

ANSWERS

of

MY LIFE

抗拒

The
ANSWERS
of
MY LIFE

不值得斗争

The ANSWERS *of* MY LIFE

玩得开心就好

The
ANSWERS
of
MY LIFE

毋庸置疑

The
ANSWERS
of
MY LIFE

你也许会失望

The
ANSWERS
of
MY LIFE

去改变

The ANSWERS *of* MY LIFE

一个强有力的承诺将会换回更好的结果

The

ANSWERS

of

MY LIFE

也许有更好的解决方案

The

ANSWERS

of

MY LIFE

不要害怕

The
ANSWERS
of
MY LIFE

想法太多，选择太少

The

ANSWERS

of

MY LIFE

是的

The
ANSWERS
of
MY LIFE

一笑而过

The
ANSWERS
of
MY LIFE

取决于你的选择

The

ANSWERS

of

MY LIFE

随他去

The

ANSWERS

of

MY LIFE

你需要考虑其他方面

The
ANSWERS
of
MY LIFE

一年后就不那么重要了

The

ANSWERS

of

MY LIFE

醒醒吧，别做梦了

The ANSWERS *of* MY LIFE

意义非凡

The

ANSWERS

of

MY LIFE

默数十秒再问我

The
ANSWERS
of
MY LIFE

去行动

The

ANSWERS

of

MY LIFE

发挥你的想象力

The

ANSWERS

of

MY LIFE

对的

The ANSWERS *of* MY LIFE

为了确保最好的决定，保持冷静

The

ANSWERS

of

MY LIFE

等待

The

ANSWERS

of

MY LIFE

你必须弥补这个缺点

The ANSWERS *of* MY LIFE

你会后悔的

The ANSWERS *of* MY LIFE

毫无疑问

The
ANSWERS
of
MY LIFE

当然

The

ANSWERS

of

MY LIFE

现在比以往任何时候的情况都要好

The

ANSWERS

of

MY LIFE

图书在版编目（CIP）数据

我的人生解答书 / 漫娱编著 .
—武汉：长江出版社，2016.4
ISBN 978-7-5492-4186-6
Ⅰ. ①我… Ⅱ. ①漫… Ⅲ. ①人生哲学 - 通俗读物 Ⅳ. ① B821-49
中国版本图书馆 CIP 数据核字（2016）第 080118 号

我的人生解答书 / 漫娱 编著

出　版	长江出版社 （武汉市解放大道 1863 号 邮政编码：430010）
出　品	漫娱文化 （湖北省武汉市积玉桥万达写字楼 11 号楼 19 层 邮政编码：430060）
出 版 人	别道玉
选题策划	长江出版社青春动漫编辑室
市场发行	长江出版社发行部
E-mail	cjpub@vip.sina.com
责任编辑	张艳艳　钟一丹
装帧设计	张　晗
印　刷	中印南方印刷有限公司
版　次	2016 年 04 月第 1 版
印　次	2024 年 10 月第 3 次印刷
开　本	780mm × 1250mm 特规 1 / 32
印　张	22
书　号	ISBN 978-7-5492-4186-6
定　价	35.00 元

电话：027-82926557(总编室) 027-82926806（市场营销部）